THE HIDDEN FIELD

IN RELATIVITY

FOUND BY A TWO DECADE SEARCH

INSPIRED BY

RICHARD P. FEYNMAN

Richard O. Calkins

ISBN: 978-1-917095-99-0

Library of Congress Cataloging-in-Publication Data:

Calkins Publishing Company LLC

rocalkins@msn.com

(425)836-3548

2125 Sahalee Dr. E. Sammamish, WA 98074 United States

This report is tendered in memory of Dr. Richard P. Feynman, one of the most brilliant and open-minded physicists of his time. A master of both humor and philosophy; he was a man for all seasons.

Even though I never knew him personally, I greatly admired him and profoundly miss him. I wish he were here to enjoy this confirmation of his belief that science must always be approached with an open mind.

Table of Contents

1 Introduction

THIS IS A CAUTIONARY TALE ABOUT THE PERILS OF
EXPERIMENTAL DESIGN

Experiments are made to examine phenomena we do not understand. Thus, there is no means by which we can know precisely how to detect what we are looking for. Whether the experiments we design to do that will produce the information we need to correctly understand it is unknowable.

There is no means by which we can determine the significance of what has not been disclosed. Thus, it is not possible to produce an experimental design that can prove, beyond any doubt, that something is true.

That is why we must leave some room for doubt even about the most thoroughly tested and validated theories in our arsenal. In the words of Richard P. Feynman:

"It is necessary and true that all of the things we say in science, all of the conclusions, are uncertain, because they are only conclusions. They are guesses as to what is going to happen, and you cannot know what will happen, because you have not made the most complete experiments."[1]

And, indeed, that is precisely what you will find in this analysis.

---

[1] Richard P. Feynman, *The Meaning of it All, Thoughts of a Citizen-Scientist (*Reading, MA: Perseus Books, 1998), 26.

2 Definitions of terms and symbols

TYPES OF MOTION AND REFERENCE FRAMES

This analysis deals with two kinds of motion and with two kinds of reference frames.

The two kinds of motion are inertial and non-inertial. Anything whose motion is **not changing** (i.e., is motionless or moving at a fixed speed in a straight line) is in **inertial** motion. Anything whose motion is **changing** is in **non-inertial** motion.

A reference frame is a physical place where experiments can be conducted and observations can be made. A reference frame whose motion is **not changing** is called an **inertial** reference frame. A reference frame whose motion is **changing** is called a **non-inertial** reference frame.

SOME IMPORTANT SYMBOLS

This symbol shows an experiment being observed by an observer in the same inertial reference frame. The ball drops vertically and is accelerated by gravity.

This symbol shows an experiment being observed by an observer in a different inertial reference frame, relative to which the experiment is traveling horizontally at V_d.

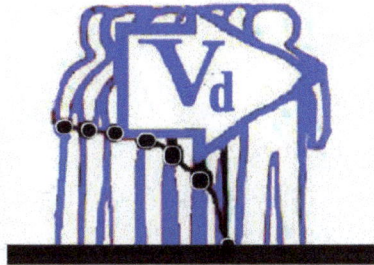

The curve in the ball's trajectory is due to gravitational acceleration.

TWO BASES USED TO MEASURE VELOCITY

V_d denotes the **difference** in velocity (aka **relative velocity**) **between different inertial reference frames**.

V_E is the velocity of an **experiment** relative to the **observer** who observes and measures it.

Note that the **same experiment** will be **observed** as being **motionless** when in the **same** reference frame as its observer ($V_E=0$),* and will be **observed** as being **in motion** at the **difference** in velocity (V_d) **between** the two reference frames when in the **other** inertial reference frame from its observer.**

Note also that the two observers are looking in opposite directions. Thus, if observer A sees inertial reference frame B as moving from left to right at velocity V_d, observer B will see inertial reference frame A as moving from left to right at velocity V_d.

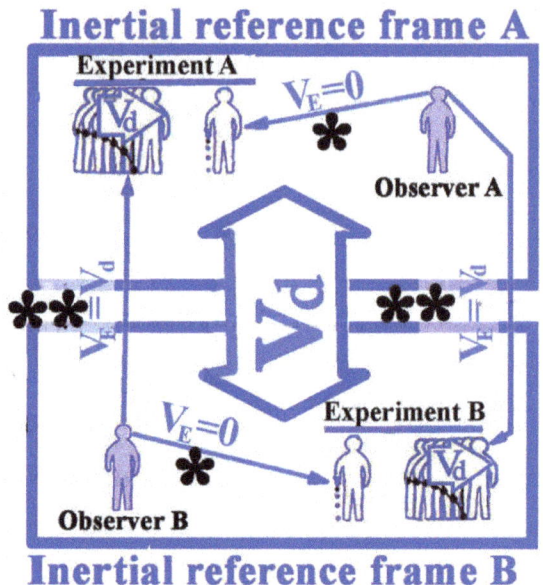

Inertial reference frame A

Experiment A

$V_E=0$

Observer A

$V_E=V_d$ V_d $V_E=V_d$

Experiment B

$V_E=0$

Observer B

Inertial reference frame B

4

3 Adherence to generally accepted theory

There are no flights of theoretical fancy in this analysis. It is based entirely on cited sources of generally accepted theory.

What this analysis does is make a simple experiment which, for a very good reason, has never been done before. It simply **changes** the motion of an experiment and its observer from one **inertial** reference frame to another. It then compares what **Newton's laws of motion** say will happen in the **non-inertial** reference frame, where their **change in motion occurs**, with what the **Galilean relativity principle** says will be **observed** in the **post-change inertial** reference frame.

Surprisingly, what this experiment reveals is that the two **disagree**! And since Einstein's first postulate of relativity, which is the foundation stone of all relativity theory, is based on the premise that they **do agree**,[2] that finding reveals that the first postulate is invalid!

One might reasonably ask, How can that be? How could it have remained undiscovered for more than a century of empirical analysis?

The purpose of this analysis is to answer those two questions and to do so based entirely on generally accepted theory. The answers are quite simple; but, as often happens, simple things can be difficult to explain.

The entire problem is a result of what seemed, at the time, to be a very reasonable decision. The phenomena observed by observers in **non-inertial** reference frames are very complex. Accordingly, Einstein decided to base his beginning effort for developing relativity theory exclusively on observations made by observers in **inertial** reference frames. Since it is limited in scope, he named it the "special theory of relativity."[3] He

---

[2] Douglas C. Giancoli, *Physics, 4th edition* (Englewood Cliffs, NJ: Prentice Hall, 1995) 742-750.

[3] Douglas C. Giancoli, *Physics*, 743.

deferred addressing the complexity of observations made by observers in **non-inertial** reference frames to his "general theory of relativity."[4]

Thus, when an experiment and its observer have their motion **changed** to move them from one **inertial** reference frame to another, the effect of that **change** is **not** determined by observing what happens in the **non-inertial** reference frame where the **change** in motion **occurs**. It is determined by the **difference** between what the observer observes in the **post-change** inertial reference frame and what he observes in the **pre-change** inertial reference frame.

Quite reasonably, the **difference** in motion between the two inertial reference frames should be the same as the **change** in motion that caused it. But, unfortunately, there are two enormous flies in the ointment.

The first enormous fly is that the **method** used to identify what is **observed** by the observer in the post-change inertial reference frame is identical to the experimental design that Galileo used to create the **Galilean relativity principle**. Thus, all it can prove is that the Galilean relativity principle agrees with itself.

The second enormous fly in the ointment is the fact that the **unanswered** question is: Does what is **observed** by the observer, after arriving in the **post-change inertial** reference frame, agree with what Newton's laws say happened in the **non-inertial** reference frame where the **change** in motion **occurred**? That requires **observing** what happened in that **non-inertial** reference frame while the **change** was in progress. And that, in turn, requires having a **non-inertial** reference frame in the experimental design. But given that the special theory deals exclusively with observations made by observers in **inertial** reference frames, that experimental design has never been used and, accordingly, that question has never even been **asked**, let alone answered.

Until now, in this analysis. It is only by asking and answering that question that one can reveal the flaw in Einstein's first postulate of relatively.

---

[4] Ibid.

6

And that requires having a **non-inertial** reference frame in the experimental design.

For purposes of clarity, throughout the rest of this analysis, material that addresses changes in motion will be shown in red, and material that addresses differences in motion (aka relative motion) will be shown in blue.

This is because the purpose of this analysis is to **compare** how **generally accepted theory** treats a change in motion (according to Newton's laws of motion) with how it treats the resulting difference in motion (according to Galileo's relativity principle), which is caused by that change.

And to do that, one must use an experimental design that provides observations by observers in **both** inertial and non-inertial reference frames.

4 The Galilean relativity principle

We will begin the analysis by examining the experimental design and its observations used by Galileo to create the Galilean relativity principle.[5]

First, we will note that it uses only observations made by observers in inertial reference frames.

Second, we will examine what the observers observe.

Third, we will examine how Galileo, and everyone since Galileo, have interpreted those observations.

---

[5] Dr. Donald Goldsmith and Robert Libbon, *Einstein: A Relative History* (New York, NY: Simon & Schuster, Inc, 2005) 67-70. See also Giancoli, *Physics*, 743-745.

GALILEO'S RELATIVITY PRINCIPLE

OBSERVATIONS

A ship is moving to the right at V_d relative to the shore. Experiments are being conducted on the ship and on the shore, which consists of an experimenter dropping a ball. An observer on the ship and one on the shore each observe both of the experiments.

Terms: V_E is the relative velocity between an **experiment** and its observer.

V_d is the relative velocity (aka, **difference** in velocity) between the two inertial reference frames.

**Same experiment, different points of view.

Observations:

- Each observer feels motionless.
- Each experiment is motionless relative to the observer in the same inertial reference frame (when $V_E=0$, ball falls vertically).
- Each observer sees the experiment in the **other** inertial reference frame as moving at $V_E=V_d$. (Both balls fall on the same curved trajectory.)

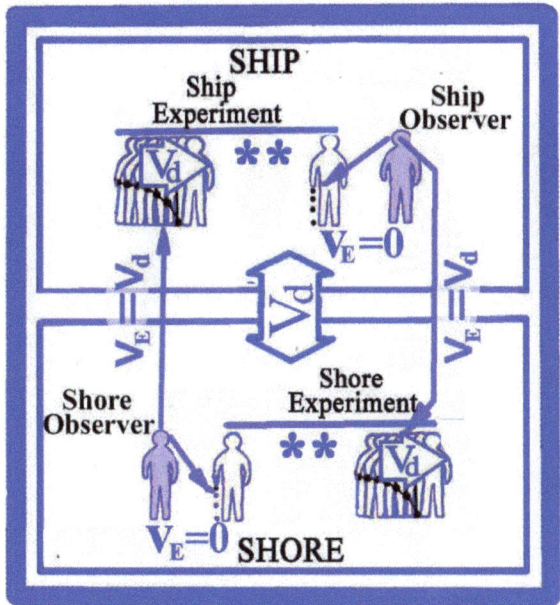

Galilean experimental design.
Inertial reference frames.

GALILEO'S RELATIVITY PRINCIPLE

INTERPRETATION OF OBSERVATIONS

It is important to note that the Galilean experimental design consists of only two **inertial** reference frames moving at a constant velocity (e.g., V_d) relative to each other.

- By definition, an observer is motionless relative to his own inertial reference frame.
- An experiment's results will be the same in every inertial reference frame (i.e., when $V_E=0$).
- The difference in motion (V_d, aka relative motion) between inertial reference frames works the same way in both directions.
- Motion can be defined only relative to the reference frame from which it is observed and measured.

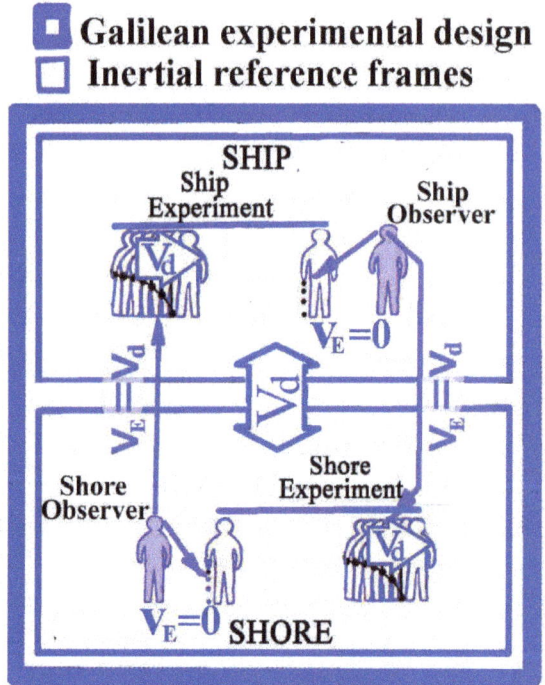

Galilean experimental design
Inertial reference frames

GALILEO'S RELATIVITY PRINCIPLE

ONE SET OF OBSERVATIONS

TWO INTERPRETATIONS

GALILEO'S OBSERVATIONS[6]

- An observer will feel motionless in every inertial reference frame.
- An experiment will produce the same result in every inertial reference frame.
- Thus, he has no means by which he can determine which inertial reference frame he is in.
- Thus, motion can be **defined** only **relative** to the reference frame from which it is observed and measured.

INTERPRETATION USED IN THE SPECIAL THEORY

- Galileo's observations define the innate characteristics of motion observed by observers in inertial reference frames. Thus, what they observe is what actually happens.

THE CORRECT INTERPRETATION WHICH IS DISCLOSED IN THIS ANALYSIS

- Because every inertial reference frame moves at a different velocity from each of the others and an observer has no means by which he can determine which one he is in, he hasn't a clue as to the speed and direction of the reference frame he uses to define the motion of what he observes.
- Motion defined relative to a frame of reference whose own motion is undefinable also is undefinable and, thus, is meaningless.

---

[6] Goldsmith, *Einstein: A Relative History*, 68-70. See also Giancoli, *Physics*, 743-745.

5 Newton's Laws of Motion

Now let's examine what generally accepted theory tells us about Newton's first two laws of motion.

This will equip us to determine what Newton's laws say will happen in the non-inertial reference frame where the change in the experiment's and observer's motion actually occurs.

Then we will be able to compare what Newton's laws say will happen in the non-inertial reference frame with what the Galilean relativity principle says will be observed in the post-change inertial reference frame.

And that, in turn, will disclose quite clearly whether or not the two agree.

Newton's first and second laws both encounter a pesky problem called a "net force." When several forces are acting on an object at different angles and different magnitudes, their interactions can be very complex. To keep it simple, Newton took the vector sum of the several forces $\Sigma\mathbf{F}$, which would have the same effect as if they were a single force $\mathbf{F}$ being applied.

Thus, a single unopposed force and a net force are essentially the same.

Multiple forces

**Vector sum
(aka, net force)
Has same effect as
the multiple forces.**

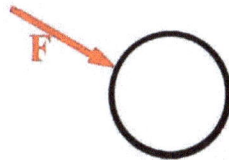

**Single unopposed
force equal to net force**

"Every body continues in its state of rest or of uniform motion in a straight line unless it is compelled to change by a net force acting on it."[7]

In plain English: A physical object will remain in inertial motion unless an unopposed external force (i.e., a net force) is applied to it to compel it to change.

This also means:

- If an unopposed external force **is not** being applied to a physical object, its motion **will not change**. It will be in inertial motion. This can be abbreviated as: No force, no change.
- If an unopposed external force **is** being applied to a physical object, its motion **will change**. It will be in non-inertial (i.e., changing) motion until such time as the force is removed. (i.e., Force applied, change occurs.)
- When the force is removed, the object's motion will simply stop changing. It will return to inertial motion at the new velocity caused by the change. And because it is moving at a **different velocity**, it will be in a **different inertial reference frame** than it was before.

Generally accepted theory refers to Newton's first law as the law of inertia.[8] However, it clearly speaks as much to what causes non-inertial motion as it does to what causes inertial motion.

---

[7] Giancoli, *Physics*, 76.
[8] Ibid.

NEWTON'S SECOND LAW OF MOTION

- The rate of change of momentum is proportional to the net force ΣF applied to it.[9] See equation 1 below. (Note: the symbol for momentum is **P**. The rate of change of momentum is the ratio of the change in momentum ΔP to the change in time Δt.)

$$\frac{\Delta P}{\Delta t} = \Sigma F \quad \mathbf{1}$$

and since
P= mV,
$\Delta P = m \Delta V$:

$$\frac{m \Delta V}{\Delta t} = \Sigma F \quad \mathbf{2}$$

and solving
for ΔV
we get:

$$\Delta V = \frac{\Sigma F}{m} \Delta t \quad \mathbf{3}$$

- Equation 3 above allows us to determine what the change in an object's velocity ΔV will be when a given net force ΣF is applied to it for a given interval of time Δt. That equips us to compare how a change in an object's motion ΔV, according to Newton's laws, will compare with the difference between what is observed by observers in the pre-change and post-change inertial reference frames.
- Recall that Einstein's special theory addresses only what is observed by observers in inertial reference frames. Thus, to address what Newton's laws will tell us about what happens in the non-inertial reference frame, where the change physically occurs, requires adding a non-inertial reference frame to the Galilean experimental design.

---

[9] Giancoli, *Physics*, 167, equation 7-2.

6 The importance of observations made from non-inertial reference frames

Recall from page 5 that Einstein's special theory is based exclusively on observations made by observers in inertial reference frames. The next experiment (pages 17 through 24) shows why that limitation produces a false conclusion that Newton's laws conform to the Galilean relativity principle.

WHY EINSTEIN'S CONCLUSION WAS INCORRECT

In the pre-change inertial reference frame below left, the ship is motionless at the shore, and both experiments are motionless relative to both observers.

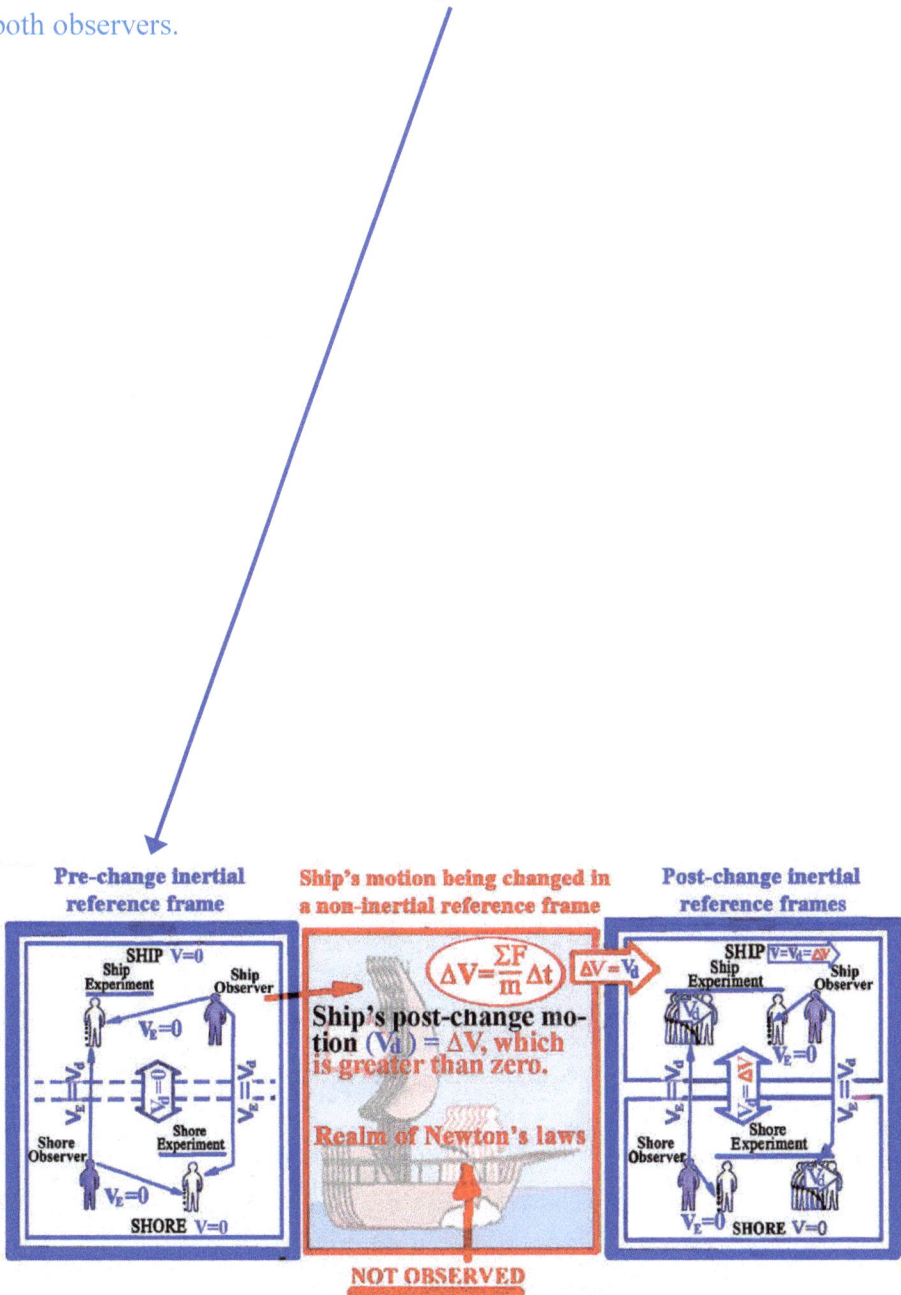

Pre-change inertial reference frame

Ship's motion being changed in a non-inertial reference frame

Post-change inertial reference frames

SHIP V=0
Ship Experiment
Ship Observer
$V_E=0$
Shore Observer
Shore Experiment
$V_E=0$
SHORE V=0

$\Delta V = \frac{\Sigma F}{m} \Delta t$ $\Delta V = V_d$

Ship's post-change motion $(V_d) = \Delta V$, which is greater than zero.

Realm of Newton's laws

SHIP V=V_d=ΔV
Ship Experiment
Ship Observer
$V_E=0$
Shore Observer
Shore Experiment
$V_E=0$ SHORE V=0

NOT OBSERVED

17

WHY EINSTEIN'S CONCLUSION WAS INCORRECT

In the pre-change inertial reference frame below left, the ship is motionless at the shore, and both experiments are motionless relative to both observers.

In the non-inertial reference frame below, the wind's force is applied to the ship's sails to change its motion. Being non-inertial, what happens there is not observed.

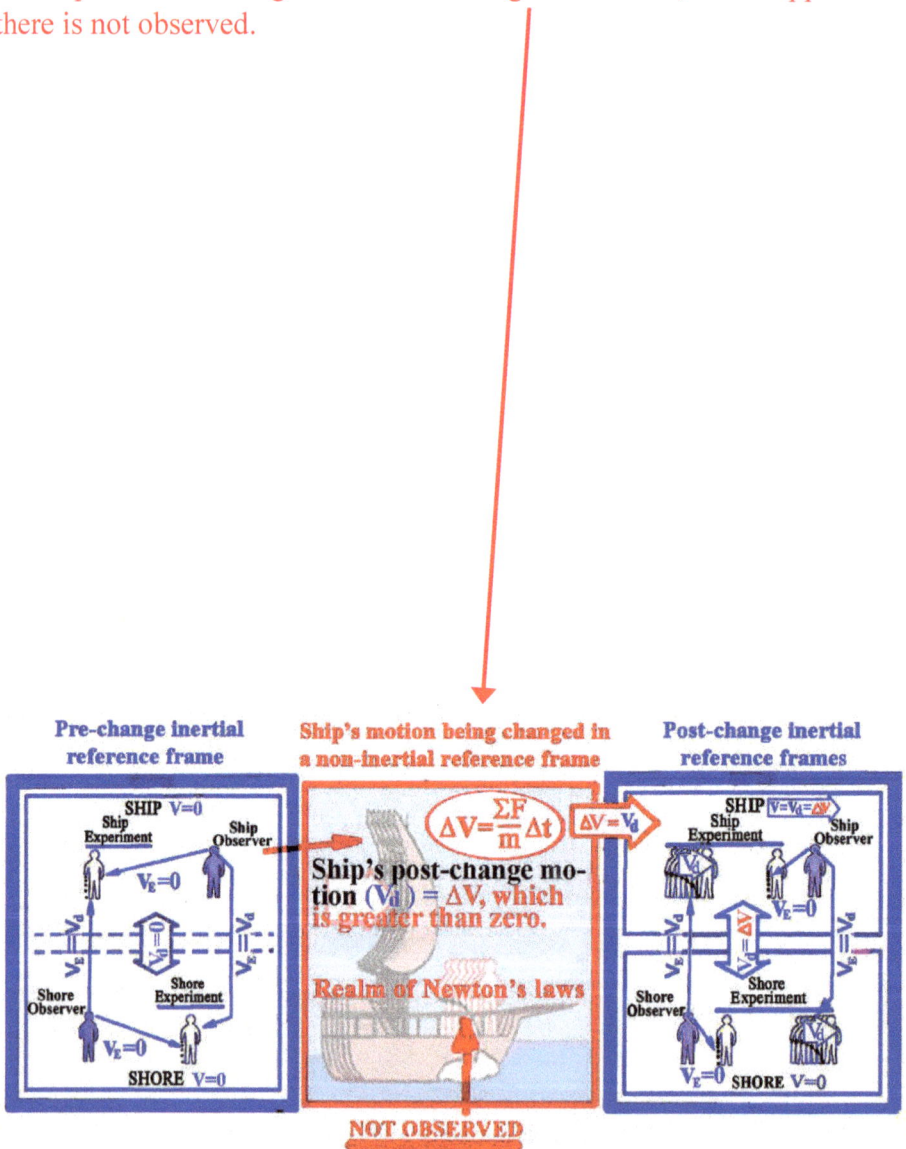

Pre-change inertial reference frame

Ship's motion being changed in a non-inertial reference frame

Post-change inertial reference frames

SHIP V=0
Ship Experiment
Ship Observer
$V_E=0$

$$\Delta V=\frac{\Sigma F}{m}\Delta t$$

$\Delta V=V_d$

Ship's post-change motion (V_d) = ΔV, which is greater than zero.

Realm of Newton's laws

Shore Observer
Shore Experiment
$V_E=0$
SHORE V=0

SHIP $V=V_d=\Delta V$
Ship Experiment
Ship Observer
$V_E=0$

$V_E=\Delta V$

Shore Observer
Shore Experiment
$V_E=0$ SHORE V=0

NOT OBSERVED

WHY EINSTEIN'S CONCLUSION WAS INCORRECT

In the pre-change inertial reference frame below left, the ship is motionless at the shore, and both experiments are motionless relative to both observers.

In the non-inertial reference frame below, the wind's force is applied to the ship's sails to change its motion. Being non-inertial, what happens there is not observed.

When the ship's change in motion is equal to V_d, the sails are furled to maintain that velocity unchanged. According to Newton's laws, changing the ship's motion will change the motion of both the **ship experiment** and the **ship observer** from $V=0$ to $V=V_d$. The change in the experiment's motion ($\Delta V = V_d$) will change its **result**. If the ball fell vertically before the change, it will fall on a curved trajectory after the change.

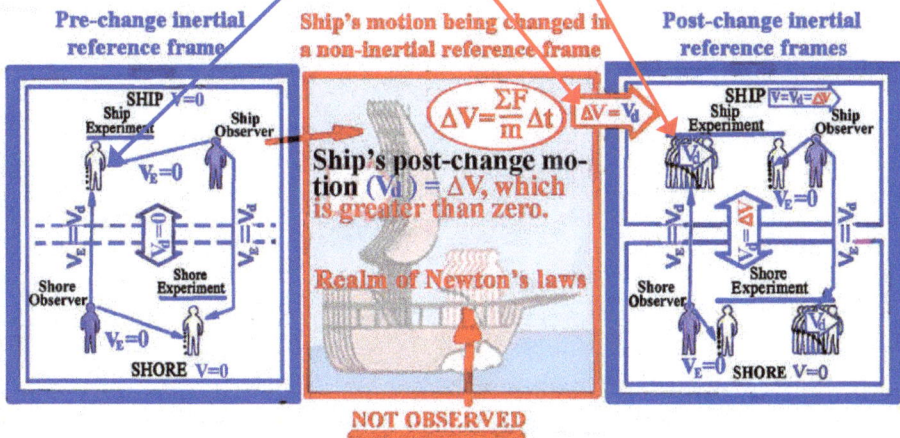

Pre-change inertial reference frame

Ship's motion being changed in a non-inertial reference frame

Post-change inertial reference frames

SHIP V=0
Ship Experiment
Ship Observer
$V_E=0$
Shore Observer
Shore Experiment
$V_E=0$
SHORE V=0

$$\Delta V = \frac{\Sigma F}{m} \Delta t$$ $\Delta V = V_d$

Ship's post-change motion (V_d) = ΔV, which is greater than zero.

Realm of Newton's laws

SHIP V=V$_d$=ΔV
Ship Experiment
Ship Observer
$V_E=0$
Shore Observer
Shore Experiment
$V_E=0$ SHORE V=0

NOT OBSERVED

In the pre-change inertial reference frame below left, the ship is motionless at the shore and both experiments are motionless relative to both observers.

In the non-inertial reference frame below, the wind's force is applied to the ship's sails to change its motion. Being non-inertial, what happens there is not observed.

When the ship's change in motion is equal to V_d, the sails are furled to maintain that velocity unchanged. According to Newton's laws, changing the ship's motion will change the motion of both the **ship experiment** and the **ship observer** from $V=0$ to $V=V_d$. The change in the experiment's motion ($\Delta V = V_d$) will change its **result**. If the ball fell vertically before the change, it will fall on a curved trajectory after the change.

However, because both the Galilean relativity principle and Einstein's special theory are based **exclusively** on observations made by observers in inertial reference frames, what happens in the non-inertial reference frame, where the change in motion physically occurs, is neither observed nor measured. The effect of the change in motion is determined by what is observed in the post-change inertial reference frame.

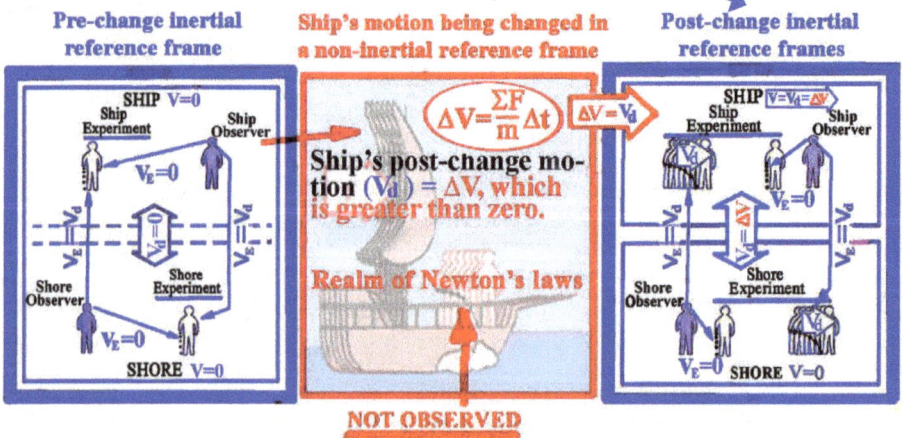

Pre-change inertial reference frame

Ship's motion being changed in a non-inertial reference frame

Post-change inertial reference frames

SHIP V=0

Ship Experiment   Ship Observer

$V_E=0$

Shore Observer   Shore Experiment

$V_E=0$

SHORE V=0

$$\Delta V = \frac{\Sigma F}{m}\Delta t$$

$\Delta V = V_d$

Ship's post-change motion (V_d) = ΔV, which is greater than zero.

Realm of Newton's laws

SHIP V=V$_d$=ΔV

Ship Experiment   Ship Observer

$V_E=0$

Shore Observer   Shore Experiment

$V_E=0$ SHORE V=0

NOT OBSERVED

20

WHY EINSTEIN'S CONCLUSION WAS INCORRECT

In the pre-change inertial reference frame below left, the ship is motionless at the shore and both experiments are motionless relative to both observers.

In the non-inertial reference frame below, the wind's force is applied to the ship's sails to change its motion. Being non-inertial, what happens there is not observed.

When the ship's change in motion is equal to V_d, the sails are furled to maintain that velocity unchanged. According to Newton's laws, changing the ship's motion will change the motion of both the **ship experiment** and the **ship observer** from $V=0$ to $V=V_d$. The change in the experiment's motion ($\Delta V = V_d$) will change its **result**. If the ball fell vertically before the change, it will fall on a curved trajectory after the change.

However, because both the Galilean relativity principle and Einstein's special theory are based **exclusively** on observations made by observers in inertial reference frames, what happens in the non-inertial reference frame, where the change in motion physically occurs, is neither observed nor measured. The effect of the change in motion is determined by what is observed in the post-change inertial reference frame.

And since the **change** took place in the **non-inertial** reference frame, which lies within the realm of Newton's laws of motion, and the **results observed** in the post-change **inertial** reference frames **conform to the Galilean relativity principle**, Einstein concluded that **Newton's laws** conform to the Galilean relativity principle.

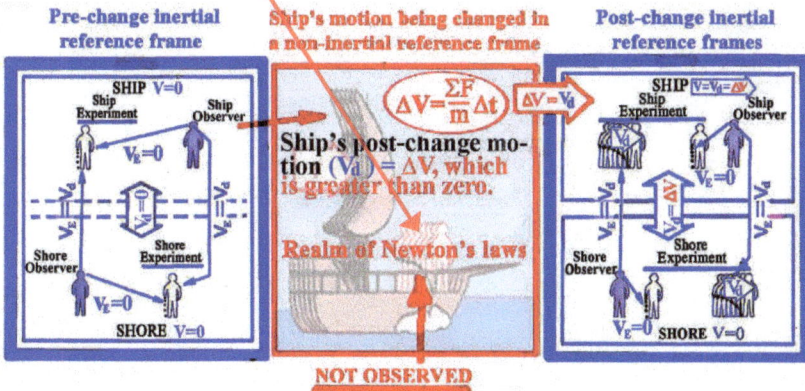

Pre-change inertial reference frame — **Ship's motion being changed in a non-inertial reference frame** — **Post-change inertial reference frames**

$$\Delta V = \frac{\Sigma F}{m} \Delta t \quad \boxed{\Delta V = V_d}$$

Ship's post-change motion (V_d) = ΔV, which is greater than zero.

Realm of Newton's laws

SHIP $V=0$
Ship Experiment — Ship Observer
$V_E = 0$

Shore Observer — Shore Experiment
$V_E = 0$
SHORE $V=0$

SHIP $V=V_d=\Delta V$
Ship Experiment — Ship Observer
$V_E = 0$

Shore Observer — Shore Experiment
$V_E = 0$
SHORE $V=0$

NOT OBSERVED

21

A CLASSIC EXAMPLE OF CIRCULAR REASONING AND AN INADEQUATE EXPERIMENTAL DESIGN

While Einstein's conclusion seems to be unquestionably correct, it is actually a very subtle example of circular reasoning applied to observations produced by an inadequate experimental design.

Note that all observations were made in pre- and post-change inertial reference frames. What actually happens in the non-inertial reference frame, where Newton's laws are in force, is not observed. But since what happened in the non-inertial reference frame produced the results observed in the post-change inertial reference frame, and the results observed conform to the Galilean relativity principle, Einstein assumed that Newton's laws must conform with the Galilean relativity principle.

Using the same observations reported in the same post-change inertial reference frame to define both cause (what must have happened in the non-inertial reference frame) and effect (what is observed in the post-change inertial reference frame) is a clear example of circular reasoning. The method used here to <u>conclude</u> that the two will agree is based on nothing more than the <u>assumption</u> that they will agree.

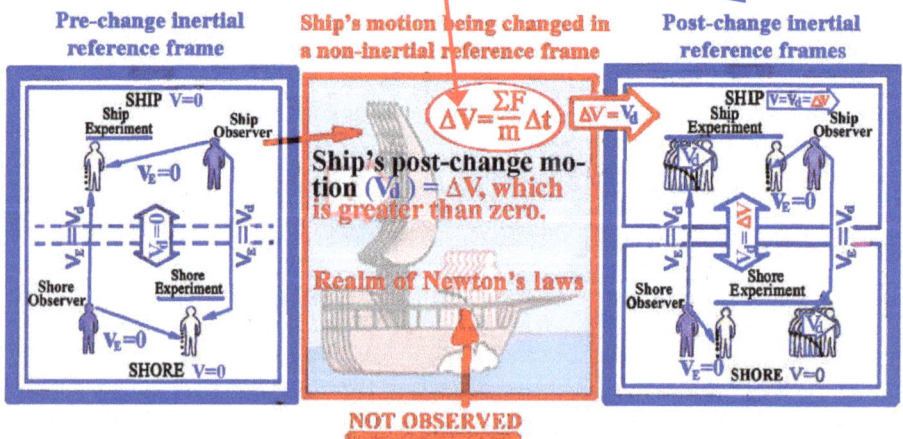

Pre-change inertial reference frame | Ship's motion being changed in a non-inertial reference frame | Post-change inertial reference frames

$$\Delta V = \frac{\Sigma F}{m} \Delta t$$

$\Delta V = V_d$

Ship's post-change motion (V_d) = ΔV, which is greater than zero.

Realm of Newton's laws

NOT OBSERVED

A CLASSIC EXAMPLE OF CIRCULAR REASONING AND AN INADEQUATE EXPERIMENTAL DESIGN

Note that if we ignore what happens in the non-inertial reference frame, we ignore the irrefutable facts that the ship observer and experiment had their motion changed (Newton's first and second laws) and the shore experiment's motion has not changed (Newton's first law). All we can know is that, in the **post-change** inertial **reference frame**, the **ship observer**, being in inertial motion, **feels motionless and since the ship experiment is motionless relative to his frame of reference ($V_E=0$), he perceives it as being motionless.** And **since the ship and shore are moving at V_d relative to each other,** he sees the shore experiment as having **changed its motion** by V_d (Galilean relativity principle). According to Newton's laws of motion, all of the ship observer's observations are incorrect.

Because it has seemed entirely reasonable to address special theory issues exclusively from the Galilean perspective (aka, using only observations made in inertial reference frames), this conflict has remained undiscovered for more than a century.

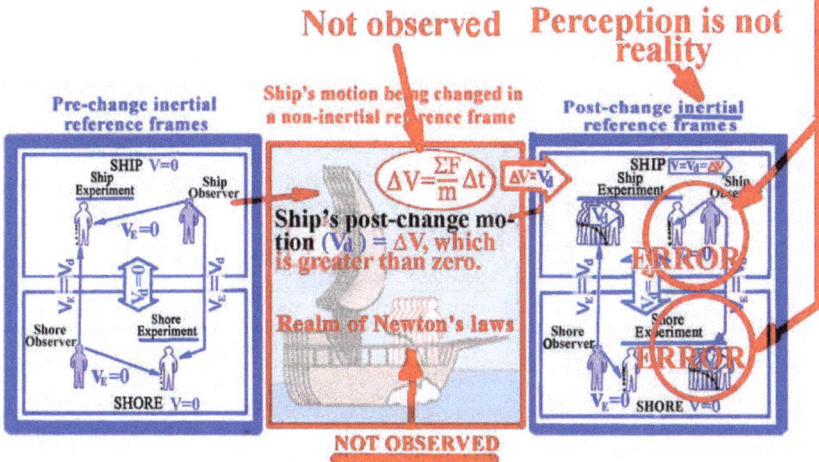

Not observed   Perception is not reality

Pre-change inertial reference frames

Ship's motion being changed in a non-inertial reference frame

Post-change inertial reference frames

SHIP V=0
Ship Experiment
Ship Observer
$V_E=0$
Shore Observer
Shore Experiment
$V_E=0$
SHORE V=0

$$\Delta V = \frac{\Sigma F}{m} \Delta t$$

$\Delta V = V_d$

Ship's post-change motion (V_d) = ΔV, which is greater than zero.

Realm of Newton's laws

SHIP V=V_d=ΔV
Ship Experiment
Ship Observer
ERROR
Shore Observer
Shore Experiment
ERROR
$V_E=0$ SHORE V=0

NOT OBSERVED

23

A CLASSIC EXAMPLE OF CIRCULAR REASONING AND AN INADEQUATE EXPERIMENTAL DESIGN

The **<u>unanswered</u>** question is: Does what happens in the non-inertial reference frame conform to what is observed in the post-change inertial reference frames? To answer that question, we must observe what happens in the non-inertial reference frame.

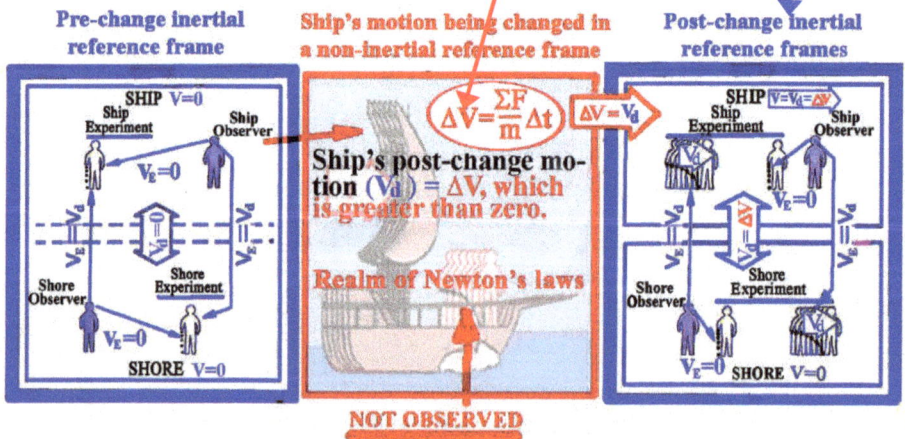

Pre-change inertial reference frame — SHIP V=0, Ship Experiment, Ship Observer, $V_E=0$, Shore Observer, Shore Experiment, $V_E=0$, SHORE V=0

Ship's motion being changed in a non-inertial reference frame — $\Delta V = \frac{\Sigma F}{m} \Delta t$, $\Delta V = V_d$, Ship's post-change motion (V_d) = ΔV, which is greater than zero. Realm of Newton's laws

Post-change inertial reference frames — SHIP V=V_d=ΔV, Ship Experiment, Ship Observer, $V_E=0$, ΔV, Shore Observer, Shore Experiment, $V_E=0$, SHORE V=0

NOT OBSERVED

7 The Newtonian experimental design

The Newtonian experimental design adds a non-inertial reference frame to the Galilean design that was used to create the Galilean relativity principle.

That reveals what Newton's laws say will happen if the ship starts out motionless at the shore and then has its motion changed to be moving at V_d relative to the shore.

That, in turn, will allow us to directly compare what Newton's laws say the change in the ship experiment's motion will be with that the Galilean relativity principle says will be observed in the post-change inertial reference frame.

- Begins with ship motionless at the shore.

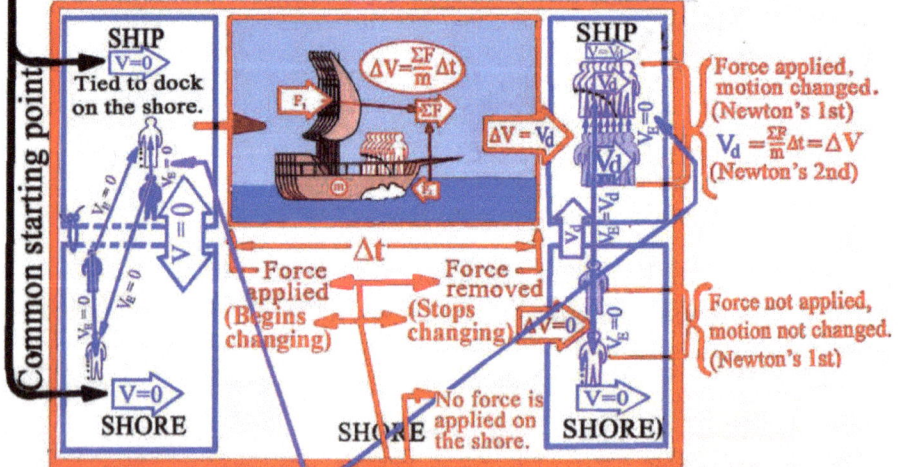

Common starting point

What both observers observe in the pre-change inertial reference frames.

What happens when an unopposed external force is applied to the ship.

What the shore observer (but not the ship observer) observes in the post-change inertial reference frames.

Pre-change inertial reference frames

Ship's motion being changed in a non-inertial reference frame.

Post-change inertial reference frames

SHIP $V=0$
Tied to dock on the shore.

$\Delta V = \frac{\Sigma F}{m} \Delta t$

ΣF

$\Delta V = V_d$

SHIP

Force applied, motion changed. (Newton's 1st)

$V_d = \frac{\Sigma F}{m} \Delta t = \Delta V$
(Newton's 2nd)

Δt

Force applied (Begins changing)

Force removed (Stops changing)

$\Delta V = 0$

Force not applied, motion not changed. (Newton's 1st)

$V=0$
SHORE

SHORE

No force is applied on the shore.

$V=0$
SHORE)

Note that $V_E=0$ on the ship both before and after its change in inertial velocity.

Ship's motion is changed.
Shore's motion is not changed.

- Begins with ship motionless at the shore.
- A force **F₁** applied to the ship will change its velocity (Newton's first).

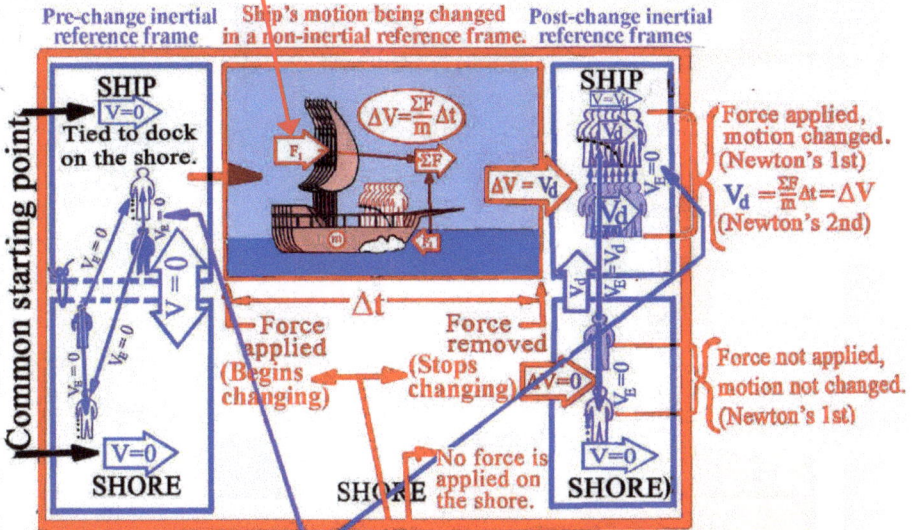

Pre-change inertial reference frame   Ship's motion being changed in a non-inertial reference frame.   Post-change inertial reference frames

SHIP
V=0
Tied to dock on the shore.

$\Delta V = \frac{\Sigma F}{m} \Delta t$

ΣF

$\Delta V = V_d$

SHIP
V=V_d

Force applied, motion changed. (Newton's 1st)

$V_d = \frac{\Sigma F}{m} \Delta t = \Delta V$ (Newton's 2nd)

Common starting point

Δt

Force applied (Begins changing)

Force removed (Stops changing)

$\Delta V=0$

Force not applied, motion not changed. (Newton's 1st)

No force is applied on the shore.

V=0
SHORE

V=0
SHORE)

SHORE

Note that $V_E=0$ on the ship both before and after its change in inertial velocity.

Ship's motion is changed.
Shore's motion is not changed.

- Begins with ship motionless at the shore.
- A force F_1 applied to the ship will change its velocity (Newton's first).
- The change in velocity ΔV will equal $\Sigma F/m \, \Delta t$ (Newton's second).

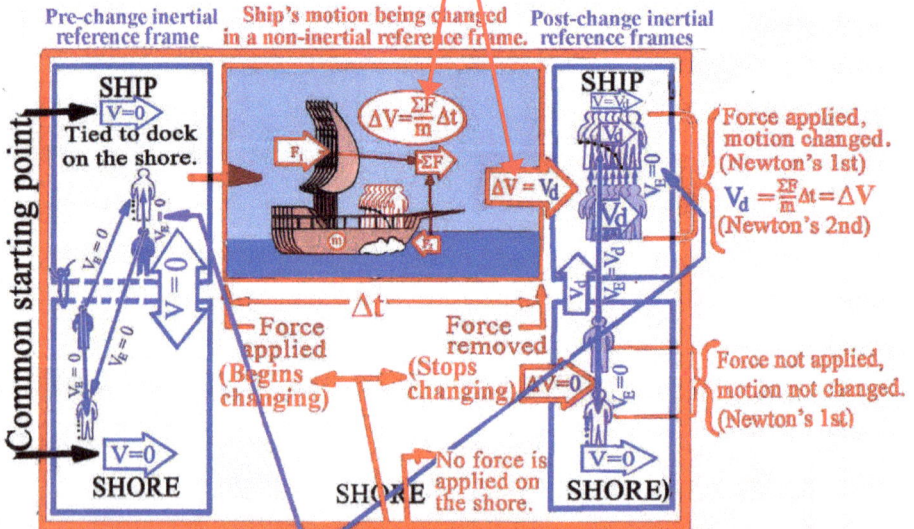

Pre-change inertial reference frame Ship's motion being changed in a non-inertial reference frame. Post-change inertial reference frames

SHIP $V=0$ Tied to dock on the shore.

$\Delta V = \frac{\Sigma F}{m} \Delta t$

$\Delta V = V_d$

SHIP $V=V_d$

Force applied, motion changed. (Newton's 1st)

$V_d = \frac{\Sigma F}{m} \Delta t = \Delta V$ (Newton's 2nd)

Common starting point

Δt

Force applied (Begins changing)

Force removed (Stops changing)

$\Delta V = 0$

Force not applied, motion not changed. (Newton's 1st)

$V=0$

SHORE

No force is applied on the shore.

$V=0$

SHORE

Note that $V_E=0$ on the ship both before and after its change in inertial velocity.

Ship's motion is changed.
Shore's motion is not changed.

THE NEWTONIAN EXPERIMENTAL DESIGN AT WORK

- Begins with ship motionless at the shore.
- A force F_1 applied to the ship will change its velocity (Newton's first).
- The change in velocity ΔV will equal $\Sigma F/m\ \Delta t$ (Newton's second).
- ΣF equals F_1 (wind on sail) minus F_2 (water's resistance).

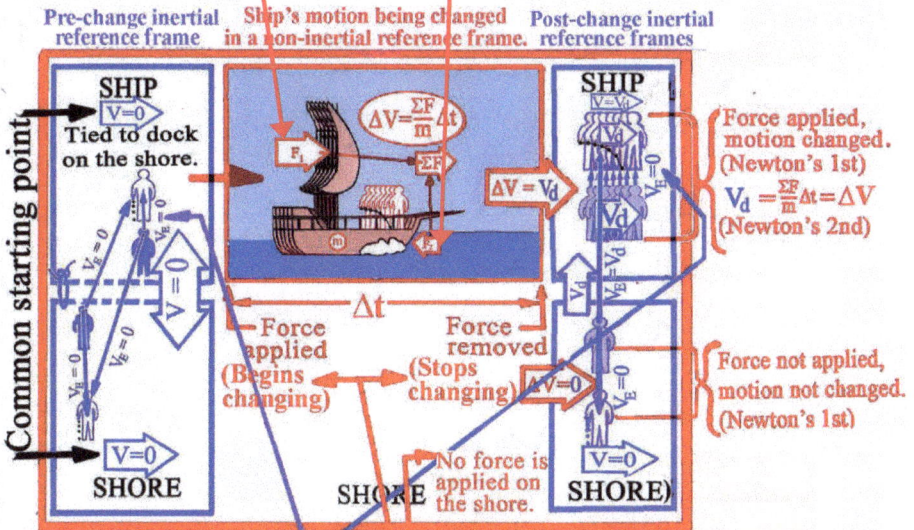

Pre-change inertial reference frame   Ship's motion being changed in a non-inertial reference frame.   Post-change inertial reference frames

Common starting point

SHIP — $V=0$ — Tied to dock on the shore.

$V_E=0$

$V=0$

SHORE

$\Delta V = \frac{\Sigma F}{m}\Delta t$

F_1

ΣF

$\Delta V = V_d$

Δt

Force applied (Begins changing)

Force removed (Stops changing)

No force is applied on the shore.

SHORE

SHIP — $V=V_d$

V_d

Force applied, motion changed. (Newton's 1st)

$V_d = \frac{\Sigma F}{m}\Delta t = \Delta V$ (Newton's 2nd)

$\Delta V=0$

Force not applied, motion not changed. (Newton's 1st)

$V=0$

SHORE

Note that $V_E=0$ on the ship both before and after its change in inertial velocity.

Ship's motion is changed.
Shore's motion is not changed.

29

- Begins with ship motionless at the shore.
- A force F_1 applied to the ship will change its velocity (Newton's first).
- The change in velocity ΔV will equal $\Sigma F/m \, \Delta t$ (Newton's second).
- ΣF equals F_1 (wind on sail) minus F_2 (water's resistance).
- When ΔV is equal to the objective value of V_d, the sail is furled so that $\Sigma F=0$.

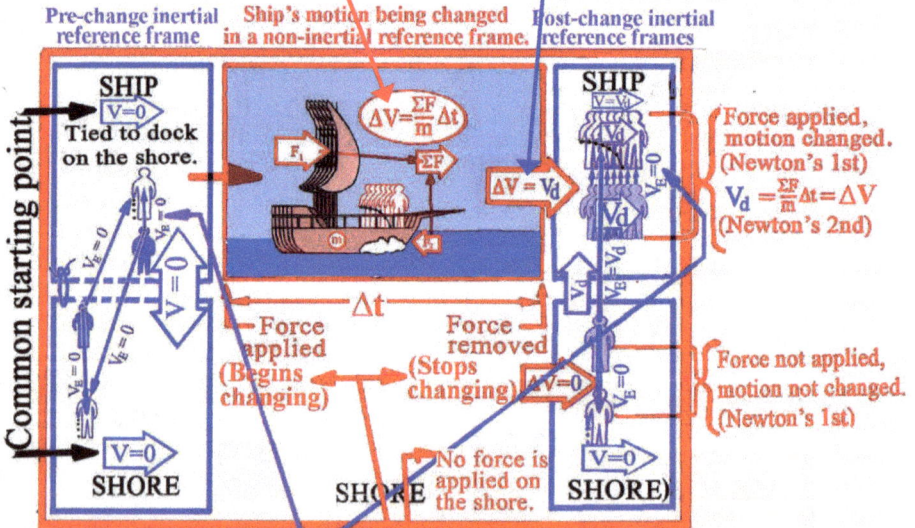

Pre-change inertial reference frame   Ship's motion being changed in a non-inertial reference frame.   Post-change inertial reference frames

SHIP
$V=0$
Tied to dock on the shore.

$\Delta V = \frac{\Sigma F}{m} \Delta t$

ΣF

F_1

F_1

$\Delta V = V_d$

V_d

SHIP
$V=V_d$

Force applied, motion changed. (Newton's 1st)

$V_d = \frac{\Sigma F}{m} \Delta t = \Delta V$
(Newton's 2nd)

Δt

Force applied (Begins changing)   Force removed (Stops changing) $\Delta V=0$

Common starting point

Force not applied, motion not changed. (Newton's 1st)

$V=0$

SHORE   **SHORE**   No force is applied on the shore. $V=0$ **SHORE**

Note that $V_E=0$ on the ship both before and after its change in inertial velocity.

Ship's motion is changed.
Shore's motion is not changed.

THE NEWTONIAN EXPERIMENTAL DESIGN AT WORK

- Begins with ship motionless at the shore.
- A force F_1 applied to the ship will change its velocity (Newton's first).
- The change in velocity ΔV will equal $\Sigma F/m\ \Delta t$ (Newton's second).
- ΣF equals F_1 (wind on sail) minus F_2 (water's resistance).
- When ΔV is equal to the objective value of V_d, the sail is furled so that $\Sigma F = 0$.
- With no net force applied to the ship, its change in momentum $\Delta P = m\ \Delta V$ will maintain its velocity at $V_d = \Delta V$ (inertial motion).

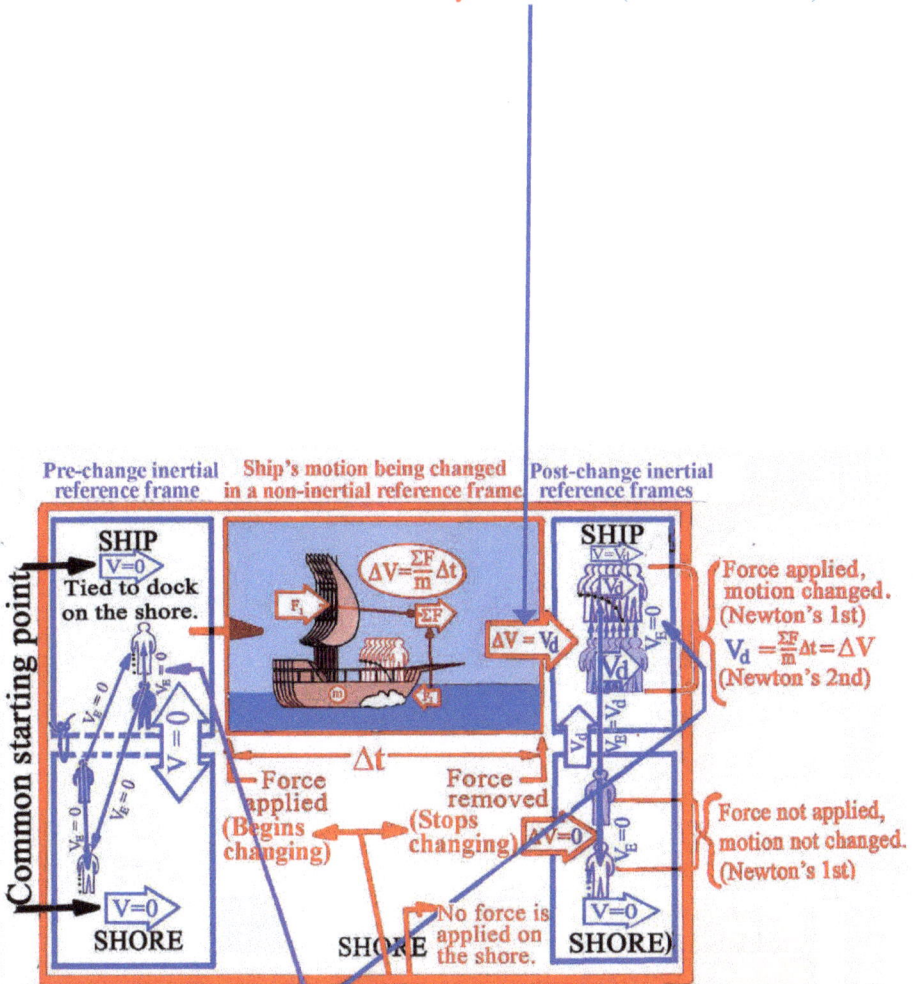

Pre-change inertial reference frame | Ship's motion being changed in a non-inertial reference frame | Post-change inertial reference frames

Common starting point

SHIP
$V=0$
Tied to dock on the shore.

$\Delta V = \frac{\Sigma F}{m} \Delta t$

ΣF

F_1

$\Delta V = V_d$

SHIP
$V = V_d$
V_d

Force applied, motion changed. (Newton's 1st)
$V_d = \frac{\Sigma F}{m} \Delta t = \Delta V$ (Newton's 2nd)

Δt

Force applied (Begins changing)

Force removed (Stops changing)

$\Delta V = 0$

Force not applied, motion not changed. (Newton's 1st)

$V=0$
SHORE

SHORE

No force is applied on the shore.

$V=0$
SHORE

Note that $V_E=0$ on the ship both before and after its change in inertial velocity.

Ship's motion is changed.
Shore's motion is not changed.

31

THE NEWTONIAN EXPERIMENTAL DESIGN AT WORK

- Begins with ship motionless at the shore.
- A force F_1 applied to the ship will change its velocity (Newton's first).
- The change in velocity ΔV will equal $\Sigma F/m \, \Delta t$ (Newton's second).
- ΣF equals F_1 (wind on sail) minus F_2 (water's resistance).
- When ΔV is equal to the objective value of V_d, the sail is furled so that $\Sigma F = 0$.
- With no net force applied to the ship, its change in momentum $\Delta P = m \, \Delta V$ will maintain its velocity at $V_d = \Delta V$ (inertial motion).
- The change in the **ship experiment's velocity** changes its result.

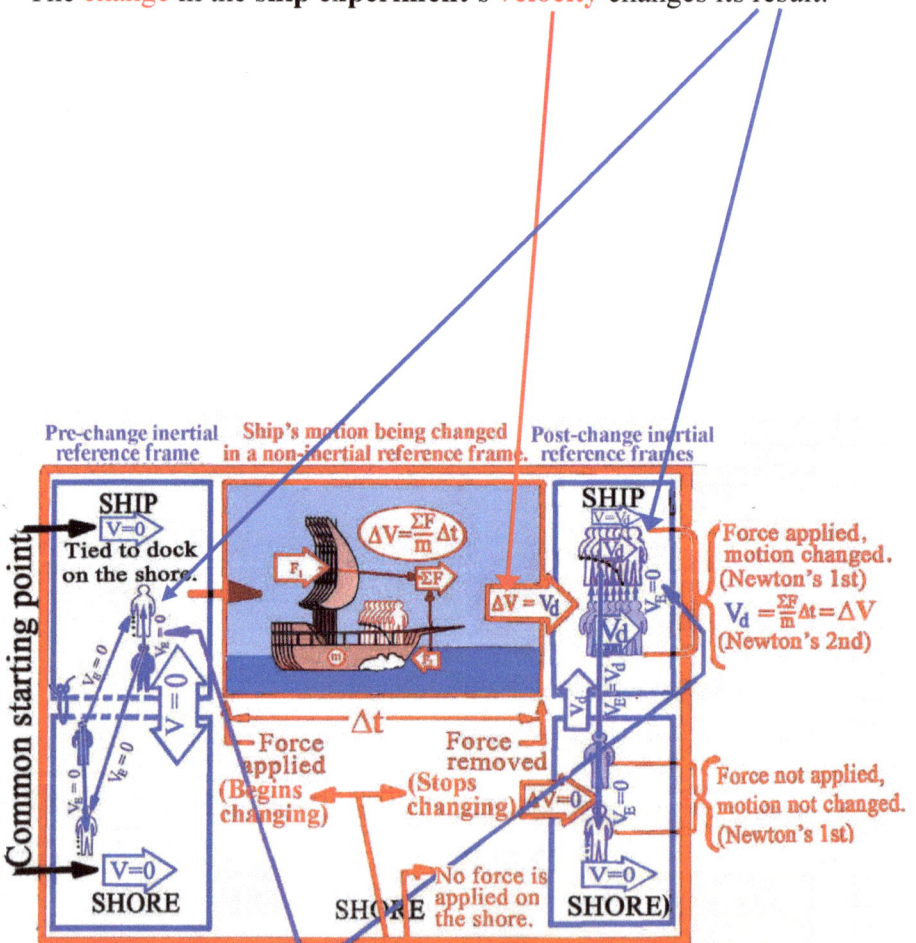

Pre-change inertial reference frame — Ship's motion being changed in a non-inertial reference frame. — Post-change inertial reference frames

SHIP $V=0$ Tied to dock on the shore.

$\Delta V = \frac{\Sigma F}{m} \Delta t$

F_1 ΣF

$\Delta V = V_d$

SHIP $V = V_d$

Force applied, motion changed. (Newton's 1st)

$V_d = \frac{\Sigma F}{m} \Delta t = \Delta V$ (Newton's 2nd)

$V_E = 0$ on the ship

Common starting point

$V = 0$

Δt

Force applied (Begins changing)   Force removed (Stops changing) $\Delta V = 0$

Force not applied, motion not changed. (Newton's 1st)

$V = 0$ SHORE   SHORE   No force is applied on the shore. $V = 0$ SHORE

Note that $V_E = 0$ on the ship both before and after its change in inertial velocity.

Ship's motion is changed.
Shore's motion is not changed.

32

- Begins with ship motionless at the shore.
- A force F_1 applied to the ship will change its velocity (Newton's first).
- The change in velocity ΔV will equal $\Sigma F/m\ \Delta t$ (Newton's second).
- ΣF equals F_1 (wind on sail) minus F_2 (water's resistance).
- When ΔV is equal to the objective value of V_d, the sail is furled so that $\Sigma F=0$.
- With no net force applied to the ship, its change in momentum $\Delta P=m\ \Delta V$ will maintain its velocity at $V_d =\Delta V$ (inertial motion).
- The change in the **ship experiment's velocity** changes its result.
- There is no change in the shore experiment's velocity.

33

WHAT THE NEWTONIAN EXPERIMENTAL DESIGN CAN DO

The Newtonian experimental design informs us that:

- Changes in motion are caused by forces acting on objects. If a force is applied to an object, its motion will change, whether or not the change is observed.[10]

- If an experiment's motion is changed, the experimental result will change, whether or not the change is observed.[11]

- If an experiment and its observer have their motion changed by the same amount, V_E will remain at zero. Thus, the observer will continue to observe the experiment as being motionless, even though its motion was changed from being motionless while in the non-inertial reference frame.

- If Newton's first and second laws of motion are valid, the Galilean relativity principle must be invalid. The two clearly do not agree.

- And since they do not agree, Einstein's first postulate of relativity is invalid.

---

[10] As shown in Chapter 4, an observer in an inertial reference frame has no means by which he can tell which inertial reference frame he is in. That is because he will feel motionless in every inertial reference frame and will see an experiment in his own inertial reference frame as being motionless.
[11] Ibid.

WHAT THE NEWTONIAN EXPERIMENTAL DESIGN CAN NOT DO

- The Newtonian design cannot reveal **why** the experimental results, caused by the change in the experiment's motion, do not conform to the relativity principle.

- The Newtonian design also fails to disclose the relationship between an observer's **perception of motion** and his own personal **state of motion** (i.e., the fact that observers in different inertial reference frames each will **perceive and define motion differently from the others**).

- There are as many different **subjective perceptions** of motion as there are observers in different inertial reference frames. Accepting each of their definitions (aka, perceptions) as having equal merit is a form of assuming that "perception is reality."[12]

---

[12] The significance of this relationship cannot be revealed by either the Galilean or the Newtonian experimental design. Only the combined Galilean/Newtonian experimental design can unveil it.

8 The Galilean/Newtonian experimental design

The Galilean/Newtonian experimental design combines the Galilean and Newtonian designs into an integrated design. What that does is reveal crucial information not provided by either the Galilean or Newtonian designs.

- It reveals **why** observations made by observers in inertial reference frames disagree with what Newton's laws say will happen when the motions of experiments and observers are changed to move them from one inertial reference frame to another.

- It reveals **how** the lack of information available to observers in inertial reference frames causes them to misinterpret what they observe.

- It reveals **why** there can be only one inertial reference frame in which observations of the results produced, even by experiments using physical objects moving at everyday speeds, will be correct. Interestingly, that is the same thing James Clerk Maxwell told us about measuring the speed of light. It can be done correctly only from one inertial reference frame. Too bad Einstein chose to trash Maxwell's unique, motionless, inertial reference frame when he adopted his second postulate of relativity.[13]

---

[13] Giancoli, *Physics*, 745-751.

The Galilean/Newtonian design **combines** the Galilean and Newtonian designs to compare how Newton's laws treat a change in motion with how Galileo's relativity principle treats the resulting difference in motion between the same two inertial reference frames.

The pre-combination building blocks are:

Newtonian Experimental Design

Changes in motion are caused by forces acting on objects and act only in the same direction as that of the applied force.

Galilean Experimental Design

Relative motion V_d between different inertial reference frames works the same way in both directions.

Effects of change in ship's motion according to Newton's laws of motion.

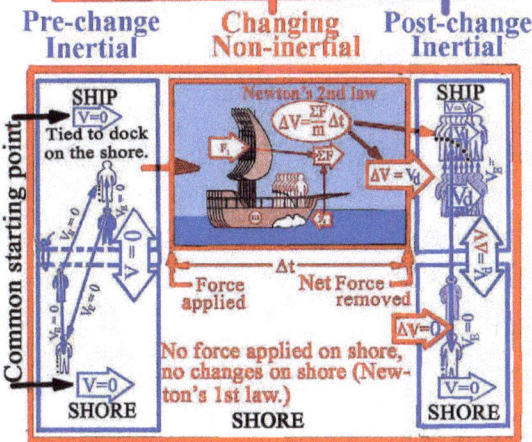

| Pre-change Inertial | Changing Non-inertial | Post-change Inertial |

Newton's 2nd law
$\Delta V = \frac{\Sigma F}{m} \Delta t$
ΣF
$\Delta V = V_d$

Force applied — Net Force removed

No force applied on shore, no changes on shore (Newton's 1st law.)

Post-change observations according to Galilean relativity principle.

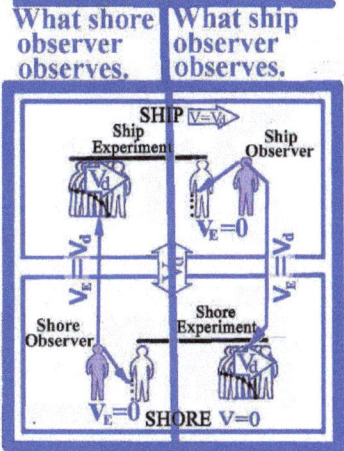

| What shore observer observes. | What ship observer observes. |

The Galilean/Newtonian design **combines** the Galilean and Newtonian designs to compare how Newton's laws treat a change in motion with how Galileo's relativity principle treats the resulting difference in motion between the same two inertial reference frames.

The shore observer's observations agree with what Newton's laws of motion say will happen when the ship's motion is changed.

Agreement

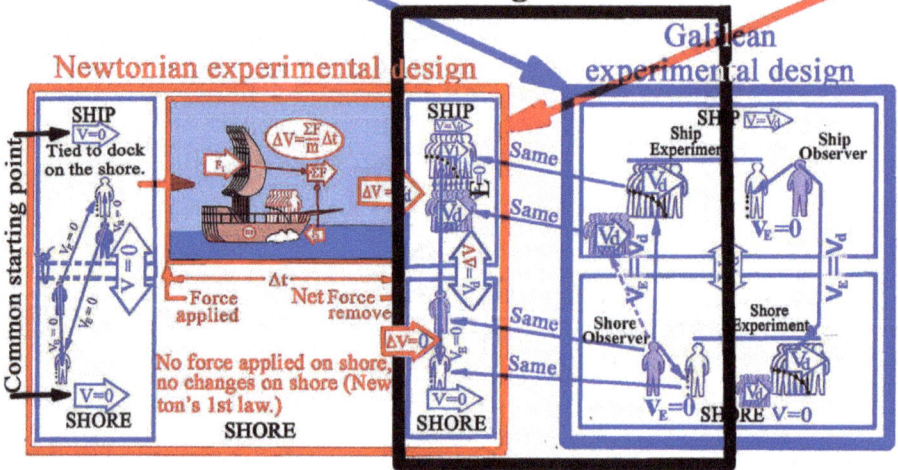

Newtonian experimental design

Galilean experimental design

The Galilean/Newtonian design **combines** the Galilean and Newtonian designs to compare how Newton's laws treat a change in motion with how Galileo's relativity principle treats the resulting difference in motion between the same two inertial reference frames.

The ship observer's observations disagree with what Newton's laws of motion say will happen when the ship's motion is changed.

Ship observer's observations

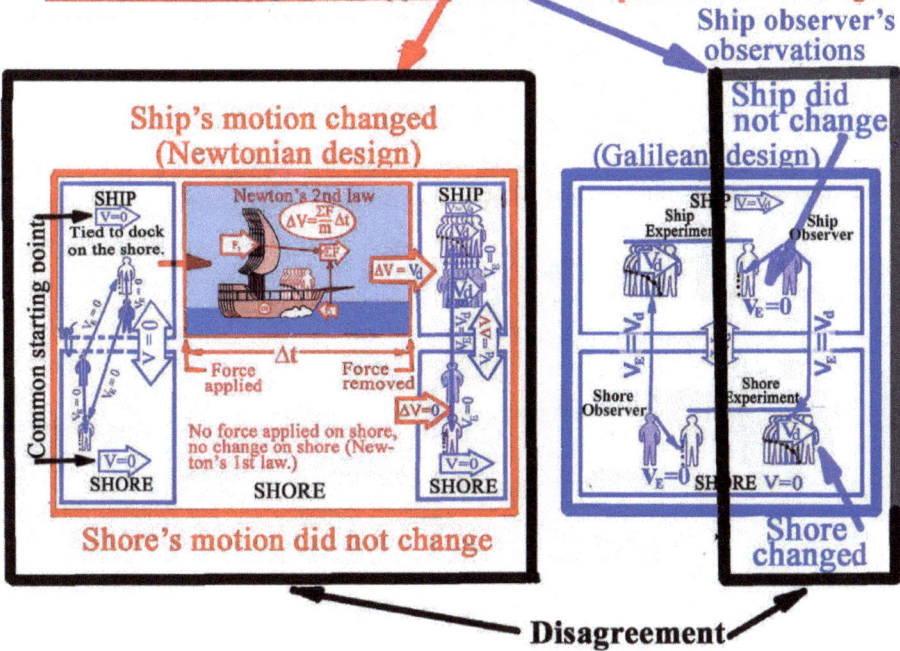

Ship's motion changed
(Newtonian design)

Ship did
not change

(Galilean design)

Shore's motion did not change

Shore
changed

Disagreement

The Galilean/Newtonian experimental design <u>overlays</u> the Galilean and Newtonian designs where they agree and leaves them separate where they disagree.

Disagreement Not overlaid **Agreement Overlaid** **Disagreement Not overlaid**

Newton and Galileo agree on what the shore observer observes.

Newtonian experimental design Changes in motion are one-way only.

They disagree on what the ship observer observes. This reveals the fatal flaw in the Galilean experimental design. Changes in motion are one-way only. But their effect on relative motion (i.e., V_d) works the same way in both directions. What this means is that after the ship is returned to inertial motion at its new velocity, the ship observer will feel motionless and will see the experiment as being motionless. He has no means by which he can determine that it is <u>his</u> motion which has changed. He believes incorrectly that it is the <u>shore's</u> motion which has changed.

SHIP
V=0
Tied to dock on the shore.

Newton's 2nd law

$\Delta V = \frac{\Sigma F}{m} \Delta t$

ΣF

$\Delta V = V_d$

SHIP
V=V_d
Ship experiment
V_d
Ship observer

Change

Δt

Force applied   Force removed

$\Delta V = 0$

Shore experiment

No change

No force applied on shore. No change on shore.

SHORE
Newton's 1st law

V=0
SHORE

A common starting condition

☐ Inertial reference frames   ☐ Non-inertial reference frame

Galilean relativity principle
Post-change observations are the same in both directions.
• **Both observers feel motionless.**
• **Each observes the experiment in his reference frame as motionless.**
• **Each observes the experiment in the other reference frame as being in motion at $V_E = V_d$.**

9 Observations made in inertial reference frames are information deficient

THIS IS WHAT NEWTON'S LAWS SAY WILL HAPPEN WHEN THE SHIP'S MOTION IS CHANGED FROM V=0 TO V=ΔV.

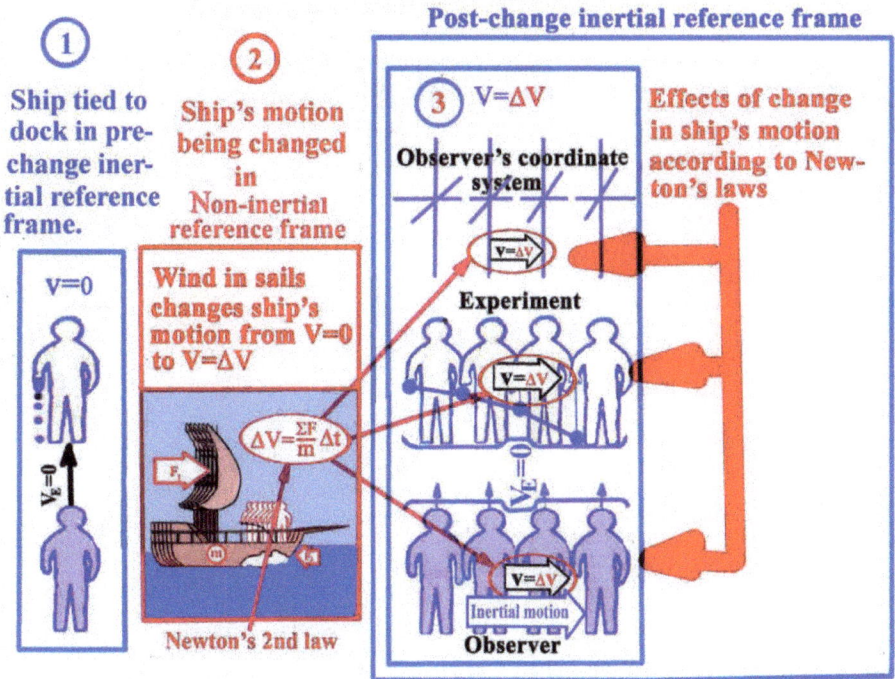

Post-change inertial reference frame

① Ship tied to dock in pre-change inertial reference frame.

② Ship's motion being changed in Non-inertial reference frame

Wind in sails changes ship's motion from V=0 to V=ΔV

$$\Delta V = \frac{\Sigma F}{m} \Delta t$$

Newton's 2nd law

③ V=ΔV
Observer's coordinate system

Experiment

V=ΔV

V=ΔV

V=ΔV
Inertial motion
Observer

Effects of change in ship's motion according to Newton's laws

WHY OBSERVATIONS MADE BY OBSERVERS IN INERTIAL REFERENCE FRAMES CANNOT PROVIDE THE INFORMATION REQUIRED TO CORRECTLY INTERPRET THEM.

If what happens to the non-inertial reference frame is not observed and measured, the only information available about the effects of the ship's change in motion is what is observed by the observer in the post-change inertial reference frame.

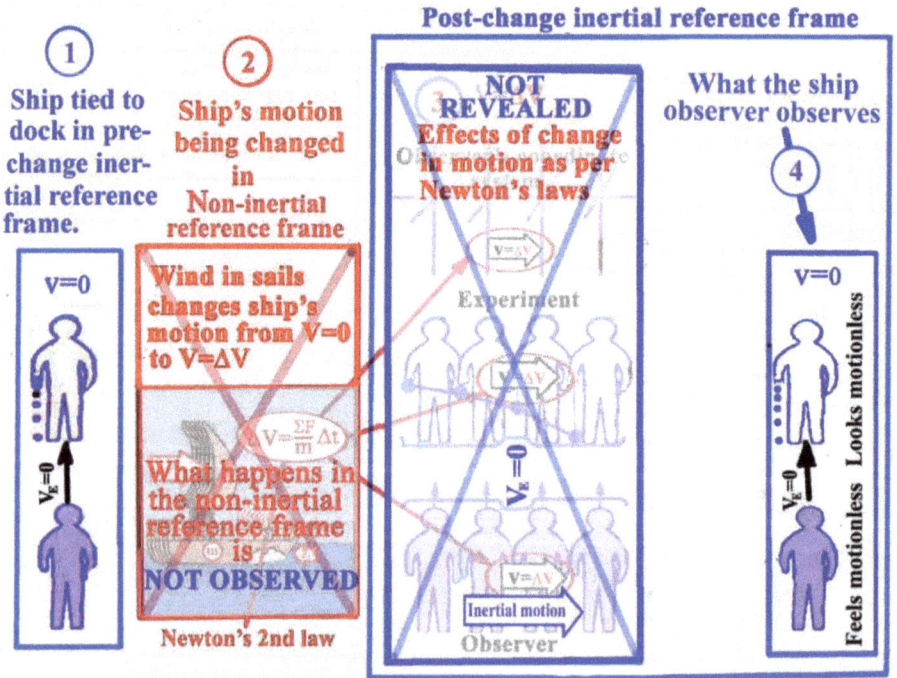

WHY OBSERVATIONS MADE BY OBSERVERS IN INERTIAL REFERENCE FRAMES CANNOT PROVIDE THE INFORMATION REQUIRED TO CORRECTLY INTERPRET THEM.

According to the Galilean relativity principle, after the ship observer arrives in the post-change inertial reference frame, he will feel motionless and see the experiment as being motionless. This is the same as what he felt and saw in the pre-change inertial reference frame. Thus, he has no means by which he can determine any difference in either the ship's or the ship experiment's motion.

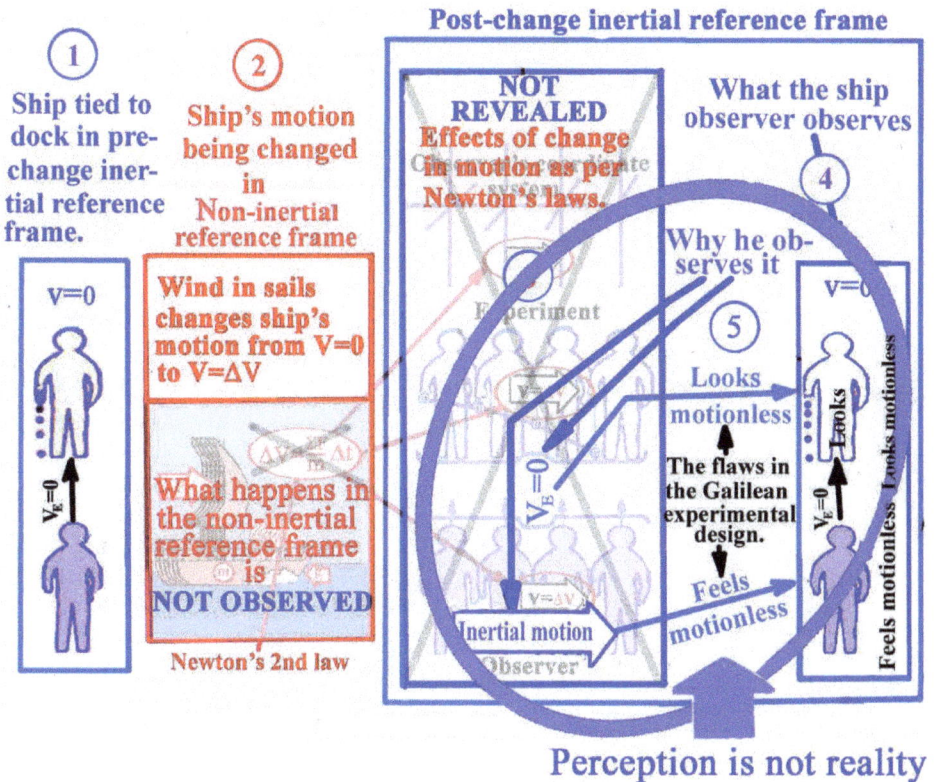

Post-change inertial reference frame

(1) **Ship tied to dock in pre-change inertial reference frame.**

(2) **Ship's motion being changed in Non-inertial reference frame**

Wind in sails changes ship's motion from V=0 to V=ΔV

What happens in the non-inertial reference frame is NOT OBSERVED

Newton's 2nd law

NOT REVEALED Effects of change in motion as per Newton's laws.

What the ship observer observes

(4)

Why he observes it

(5)

Looks motionless

The flaws in the Galilean experimental design.

Feels motionless

Looks motionless

Feels motionless

$V=0$

$V_E=0$

$V=\Delta V$

Experiment

Inertial motion Observer

Perception is not reality

43

10 The difference between Newton's laws of motion and the Galilean relativity principle

The reason why Newton's laws of motion cannot be reconciled with the Galilean relativity principle is because:

THEY ARE DIFFERENT IN KIND.

They address different aspects of motion, having different causes and different effects.

They have nothing in common.

This can be revealed only by observing what happens in a non-inertial reference frame, despite the fact that the special theory of relativity is based exclusively on observations made by observers in inertial reference frames.

THEY HAVE NOTHING IN COMMON

Different in Kind

Relative motion and Changing motion are different in kind.

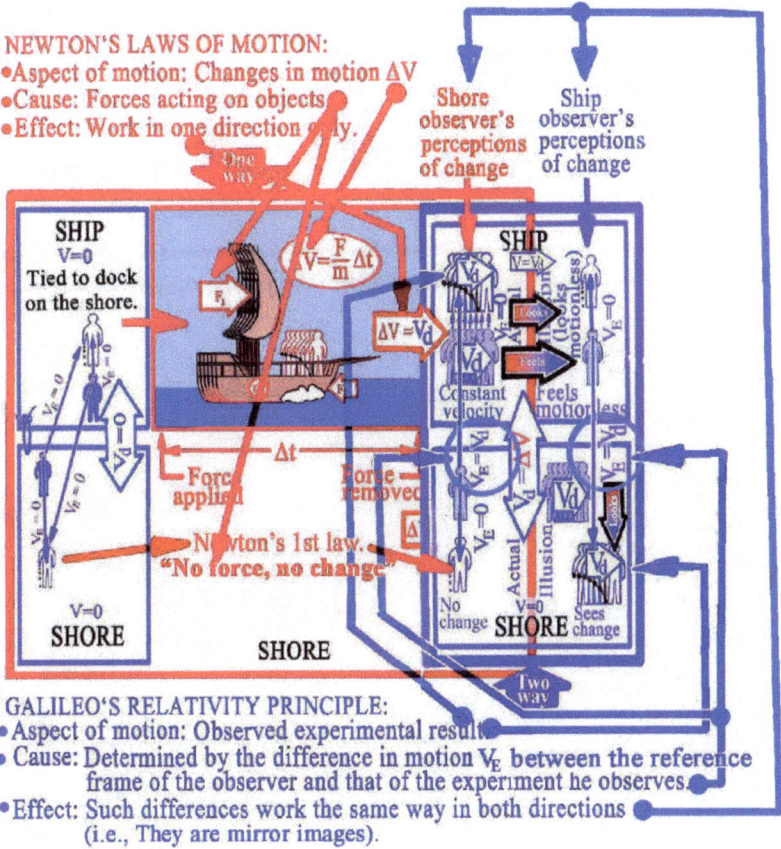

NEWTON'S LAWS OF MOTION:
- Aspect of motion: Changes in motion ΔV
- Cause: Forces acting on objects.
- Effect: Work in one direction only.

One way

Shore observer's perceptions of change

Ship observer's perceptions of change

SHIP
$V=0$
Tied to dock on the shore.

F_1

$V=\frac{F}{m}\Delta t$

$\Delta V=V_d$

Δt

Force applied

Force removed

Newton's 1st law. "No force, no change"

$V=0$

SHORE

SHORE

SHIP
$V=V_d$

V_d

V_d

Constant velocity

Feels motionless

Looks motionless

V_E

V_E

$V_E=\Delta V$

V_E

Actual

V_d

Illusion

V_d

No change

$V=0$

SHORE

Sees change

Two way

GALILEO'S RELATIVITY PRINCIPLE:
- Aspect of motion: Observed experimental result
- Cause: Determined by the difference in motion V_E between the reference frame of the observer and that of the experiment he observes.
- Effect: Such differences work the same way in both directions (i.e., They are mirror images).

THEY PRODUCE DIFFERENT RESULTS

Changes in motion (ΔV) work in one direction only.
Relative motion (V_d) works the same way in both directions.

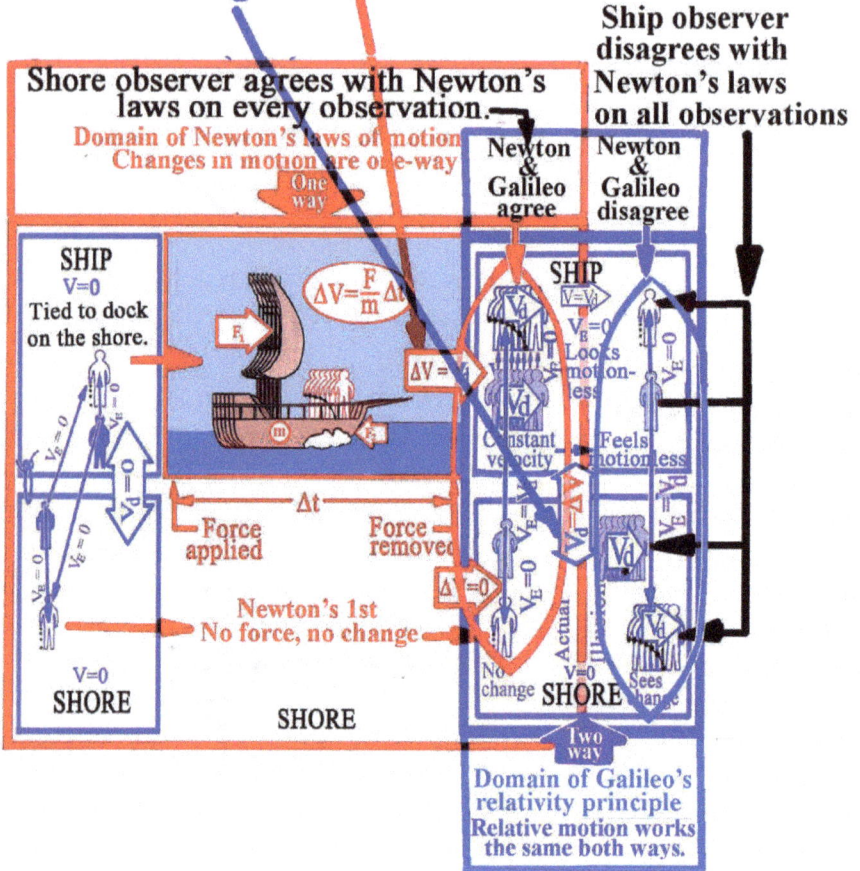

Ship observer
disagrees with
Shore observer agrees with Newton's   Newton's laws
laws on every observation.   on all observations

Domain of Newton's laws of motion
Changes in motion are one-way

One way

Newton & Galileo agree	Newton & Galileo disagree

$$\Delta V = \frac{F}{m} \Delta t$$

SHIP
V=0
Tied to dock
on the shore.

F_1

$\Delta V = $

SHIP
$V = V_d$
$V_E = 0$
Looks motion-less

V_d

V_d

Constant velocity

Feels motionless

Δt

Force applied   Force removed

$\Delta V = 0$

ΔV

Actual

V_d

Newton's 1st
No force, no change

V=0
SHORE

No change   V=0   Sees change

SHORE   SHORE

Two way

Domain of Galileo's
relativity principle
Relative motion works
the same both ways.

11 Observers in different inertial reference frames perceive and define motion differently

Every observer's definition of motion is completely subjective and is determined by his own speed and direction of motion, (i.e., is determined by which inertial reference frame he is in).

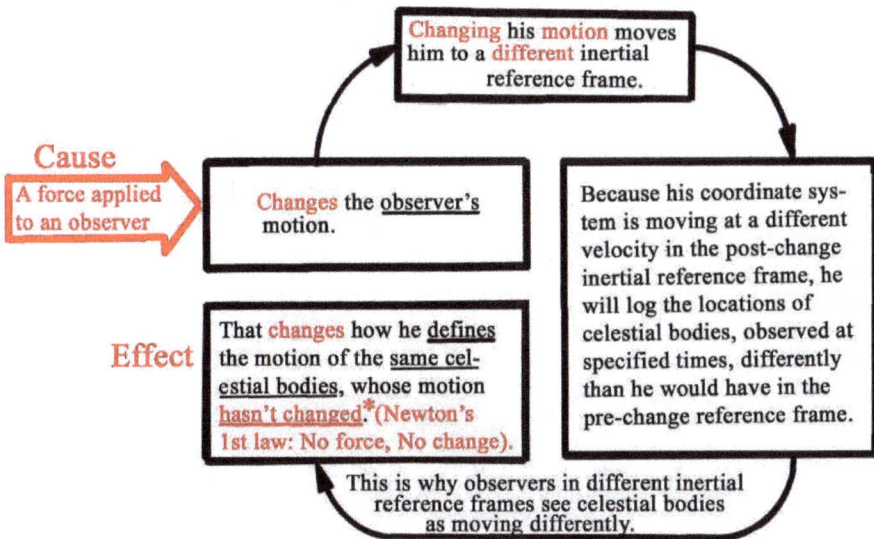

Changing his motion moves him to a different inertial reference frame.

Cause

A force applied to an observer

Changes the observer's motion.

Because his coordinate system is moving at a different velocity in the post-change inertial reference frame, he will log the locations of celestial bodies, observed at specified times, differently than he would have in the pre-change reference frame.

Effect

That changes how he defines the motion of the same celestial bodies, whose motion hasn't changed.*(Newton's 1st law: No force, No change).

This is why observers in different inertial reference frames see celestial bodies as moving differently.

* Only the observer's motion has been changed.

The example on the next three pages shows how changing an observer's motion to a different inertial reference frame gives him a completely different perception of the motion of celestial bodies.

Everything is motionless

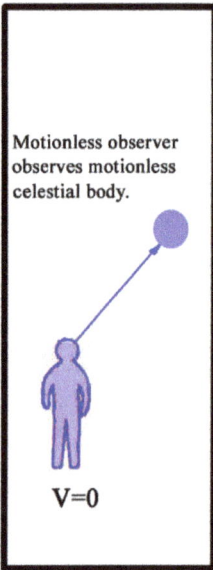

Motionless observer observes motionless celestial body.

V=0

Everything is motionless

Observer's motion is changed (what happens)

A force applied to the observer changes his horizontal velocity by ΔV.

Celestial body appears to be motionless.

$V=0$

A force is applied to the observer to change his motion from motionless to moving to the right at $V=\Delta V$. This is what will happen according to Newton's laws.

Snapshots taken at uniformly spaced points in time.

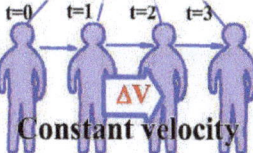

$t=0$ $t=1$ $t=2$ $t=3$

ΔV

Constant velocity

After the observer's motion has been changed, he will see celestial bodies moving differently than they did in his pre-change inertial reference frame. It is this unrecognized flaw in the Galilean experimental design that created the belief in multiple universes.

Everything is motionless	Observer's motion is changed (what happens)	What the observer will feel and see (what he perceives)

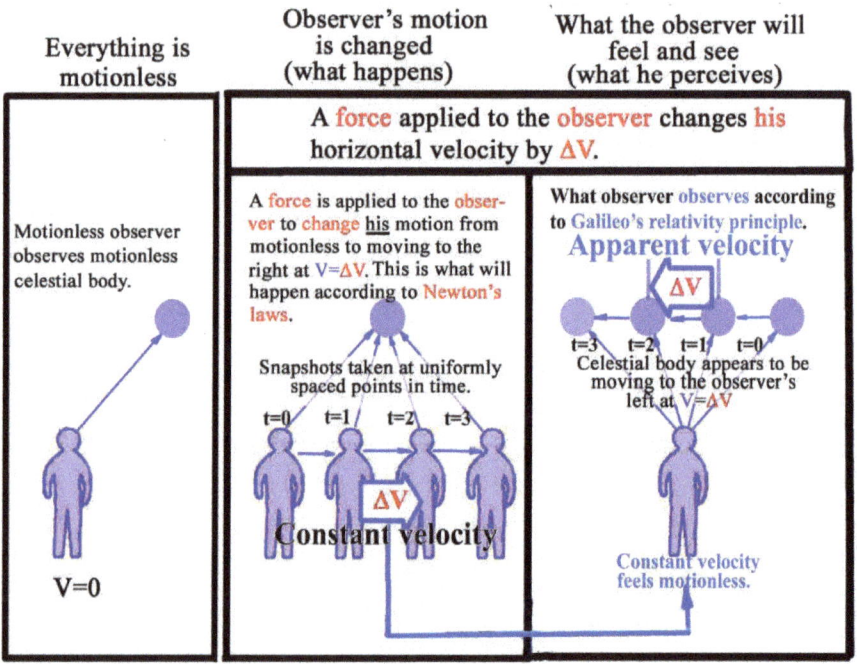

A **force** applied to the **observer** changes **his** horizontal velocity by ΔV.

Everything is motionless

Motionless observer observes motionless celestial body.

$V=0$

Observer's motion is changed (what happens)

A **force** is applied to the **observer** to **change his** motion from motionless to moving to the right at $V=\Delta V$. This is what will happen according to **Newton's laws**.

Snapshots taken at uniformly spaced points in time.

$t=0$ $t=1$ $t=2$ $t=3$

ΔV

Constant velocity

What the observer will feel and see (what he perceives)

What observer **observes** according to **Galileo's relativity principle**.

Apparent velocity

ΔV

$t=3$ $t=2$ $t=1$ $t=0$

Celestial body appears to be moving to the observer's left at $V=\Delta V$

Constant velocity feels motionless.

50

12 Conclusions

This analysis does not rely on any new theoretical concepts. All it does is disclose that what generally accepted theory tells us about Newton's laws of motion and the Galilean relativity principle, when addressed separately, directly conflicts with what it tells us about their relationship in Einstein's first postulate of relativity. This inarguably invalidates the special theory of relativity.

It is really that simple.

The flaw was created by, and subsequently has been hidden by, Einstein's very reasonable decision to base his beginning theory of relativity exclusively on observations made by observers in inertial reference frames.

Ironically, the flaw created by that decision cannot be detected by experimental designs that omit non-inertial reference frames.

Again, it is really that simple.

RELATED CONCLUSIONS

1. The special theory of relativity is rendered invalid because the first postulate of relativity is invalid. The first postulate is invalid because Galileo's relativity principle is invalid. Galileo's relativity principle is invalid because the observations upon which it is based were produced by an experimental design which did not provide the information required to correctly interpret them. (That experimental design ignores Newton's laws.)

2. As shown in equation 3 on page 15, Newton's second law produces a different experimental result in every inertial reference frame to which the experiment is moved (each different value of Δt produces a different value of ΔV which produces a different experimental result). But, being in inertial motion with $V_E=0$, an observer who accompanies the experiment will feel motionless and will observe (aka perceive) the experiment as motionless. Thus, there can be only one inertial reference frame in which the observer's observation is correct.

    Note that two completely different experimental designs, dealing with entirely different phenomena, both show that a unique, genuinely motionless, inertial reference frame exists. It should be clear, to anyone with an open mind, that Einstein's decision to reject, out of hand, Maxwell's unique motionless inertial reference frame was an incorrect, and factually unsupported, decision. It was nothing more than an act of hubris. And it is opposed not only by Maxwell's equations but by Newton's laws of motion.

    And no matter how one deals with it, the conflict between Newton's laws and the Galilean relativity principle invalidates Einstein's first postulate of relativity. It's that simple. How did we overlook it for so long? It took me two decades and I was stubbornly seeking it. After all, James Clerk Maxwell and Richard P. Feynman both told me it was there.

3. The Inarguable Relativity Killer

The reason why Newton's laws of motion and Galileo's relativity principle do not agree with each other is because they are **different in kind**.

Newton's laws deal with:
- Physical changes in motion,
- Which are caused by forces acting on objects, and
- Which operate only in the same direction as the force which caused them.

Galileo's relativity principle deals with:
- Motions of objects observed by observers in inertial reference frames,
- Which are defined relative to the reference frame from which they are observed, **whose speed and direction of motion are unknown**, and
- Which work in the same way in both directions.

They deal with different aspects of motion, having different causes and different effects. They have nothing in common.

	Newton's laws	Galileo's relativity principle
Aspect of motion	Physical changes in motion	Motions of objects as observed by observers in inertial reference frames
Cause	Forces acting on objects	Defined relative to the frame of reference from which they are observed
Effect	Operate in one direction only	Operate the same way in both directions
Validity	Defined in absolute terms	Defined relative to reference frames whose motion is unknown

4. There are two ways to deal with the genuine difference in kind between Newton's laws of motion and Galileo's relativity principle. One way is to correct the observations (aka perceptions) for the difference in motion between the observer's frame of reference and the one in which observations will be correct. That is the method shown to be necessary by James Clerk Maxwell when he solved the mysteries of the propagation of light.[14]

The other way is to make relativistic adjustments to the parameters of the equations of Newton's laws (i.e., the properties of time, space, and mass)[15] to make observations conform to Galileo's relativity principle. This is the method chosen by Einstein.

Maxwell's method corrects perceptions to discern reality. Ironically, Einstein's method has the effect of adjusting reality (i.e., the characteristics of time, space, and mass) to conform to perception. That is what causes the special theory's magical predictions.

---

[14] Giancoli, *Physics*, 745-746.

[15] Ironically, these are as much adjustments to Newton's laws as are the terms Maxwell added to their calculations. Since the first postulate holds that Newton's laws will conform to Galileo's relativity principle without need for adjustment, the special theory violates the very postulate on which it is based.

RELATED CONCLUSIONS

5. The usual defense against analyses of this type is to invoke the special theory's relativistic effects (altering time, space, and mass for relative velocity) to reconcile Newton's laws with Galileo's relativity principle. But that defense won't work here. According to generally accepted theory, such relativistic effects, even if they existed, would not be apparent at everyday speeds.[16]

    This analysis has shown that Newton's laws disagree with Galileo's relativity principle based on experiments using physical objects at everyday speeds. Thus, the Galileo/Newton relativity principle has been proven invalid under the same conditions as it was created. The proof resulted from correcting the infirmities in the experimental design on which it was based (i.e., by adding a non-inertial reference frame to the experimental design).

    Given that the Galilean relativity principle is invalid for the motion of physical objects at everyday speeds, promoting it to a universal principle for all phenomena involving motion is not going to make it valid.

---

[16] Giancoli, *Physics*, 753, 755.

IN CLOSING

Because the special theory is invalid, its predictions are unfounded. There are no multiple universes. All of the virtually infinite set of inertial reference frames coexist in and expand to the boundaries of the same universe. They differ only in the speeds and direction in which they travel, which determine the observed motion of everything observed from them.

It also follows that an object's mass does not increase to infinity as its speed approaches the speed of light. Nor does $\mathbf{E=mc^2}$. And a black hole has no more mass than the star whose supernova terminated its nuclear activity. Admittedly, absent the nuclear turmoil which expanded its use of space, it is far more tightly packed than it was before. But its mass is no greater, at least not initially.

It also is quite a bit blacker than before, given that its nuclear activity has been terminated. However, given that its charged particles are still in motion, albeit at a much-reduced level of activity, there still will be an occasional small burp of electromagnetic radiation. And because its mass is roughly the same as before, its gravitational pull will tend to collect whatever debris wanders by as effectively as it did when it was a star. But since it no longer incinerates what it collects, its mass will, indeed, grow greater than when it was a star. But it won't become infinite.

It also is not unreasonable to posit that energy and mass are different in kind and are not interchangeable. Energy is innately dynamic, and its force moves mass. Mass is innately inert, and its motion is changed, however reluctantly, by the application of forces (aka energy).

And time and space are not interchangeable. The postulate on which that depends was based on clueless interpretations of fatally incomplete observations. And just consider the implications for man in space. We now know that the limits created by Einstein's twisting of time, space, and mass, to make his postulates appear to be correct, are merely fiction. And that means spaceships can accelerate to virtually any speed. The only thing that determines the length of a trip is the time it takes to accelerate and decelerate at rates that man and the spaceship can tolerate.

AND FINALLY:

SOME THOUGHTS ON THE PERILS OF DEPENDING UNKNOWINGLY ON AN INADEQUATE EXPERIMENTAL DESIGN

"Perception is not reality."

Anonymous

"Things must be learned only to be unlearned again, or more likely to be corrected."

Richard P. Feynman, 1963

"Wisdom begins with the understanding that everything known by man is subjective. It never can be known beyond doubt."

Richard O. Calkins, 2023

SO...WHAT IS THE ANSWER?

HOW CAN WE CORRECTLY DEFINE MOTION?

Interestingly, we already have the answer. We got it from the brilliant work of James Clerk Maxwell.[17] What he told us is now backed up by learning what is presented in this analysis:

- Adding a non-inertial reference frame to the Galilean experimental design reveals that what Newton's laws say will happen when an experiment and its observer have their motion changed, to move them to a different inertial reference frame, disagrees with what the Galilean relativity principle says will be observed by the observer in the post-change inertial reference frame. That disagreement invalidates Einstein's first postulate of relativity.[18]

- According to Newton's laws, if identical experiments involving the motion of physical objects have their motion changed to move them from a common pre-change inertial reference frame to different post-change inertial reference frames, the experimental result will be different in every post-change inertial reference frame. However, due to insufficient information, the result observed by the observers in every post-change inertial reference frame will be the same. Thus, there can be only one inertial reference frame where the observed experimental result will be correct.

- And Maxwell's equations for the propagation of light tell us how to identify it.

---

[17] Giancoli, *Physics,* 742-750.
[18] Ibid.

THE END

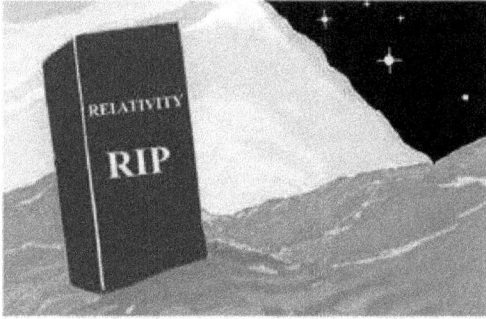

For additional information and free downloads, go to:

www.calkinspublishing.com

www.ingramcontent.com/pod-product-compliance
Lightning Source LLC
Chambersburg PA
CBHW040910210326
41597CB00029B/5030